NATURE'S GROSSEST

CATERPILLARS EAT THEIR SKIN!

By Bert Wilberforce

Please visit our website, www.garethstevens.com. For a free color catalog of all our high-quality books, call toll free 1-800-542-2595 or fax 1-877-542-2596.

Cataloging-in-Publication Data

Names: Wilberforce, Bert.
Title: Caterpillars eat their skin! / Bert Wilberforce.
Description: New York : Gareth Stevens Publishing, 2018. | Series: Nature's grossest | Includes index.
Identifiers: ISBN 9781538209370 (pbk.) | ISBN 9781538209394 (library bound) | ISBN 9781538209387 (6 pack)
Subjects: LCSH: Caterpillars–Juvenile literature. | Butterflies–Juvenile literature. | Moths–Juvenile literature.
Classification: LCC QL544.2 W57 2018 | DDC 595.78'139–dc23

Published in 2018 by
Gareth Stevens Publishing
111 East 14th Street, Suite 349
New York, NY 10003

Copyright © 2018 Gareth Stevens Publishing

Designer: Laura Bowen
Editor: Therese Shea

Photo credits: Cover, p. 1 Cathy Keifer/Shutterstock.com; pp. 3–24 (background) Oleksii Natykach/Shutterstock.com; p. 5 (butterfly) CHAINFOTO24/Shutterstock.com; p. 5 (moth) Simon Kovacic/Shutterstock.com; p. 7 Henrik Larsson/Shutterstock.com; p. 9 (top left) Mark Zulkoskey/Shutterstock.com; p. 9 (top right) Tyler Fox/Shutterstock.com; p. 9 (bottom left) Matee Nuserm/Shutterstock.com; p. 9 (bottom right) Hibiki Nakata/Shutterstock.com; p. 11 Dorling Kindersley/Getty Images; p. 13 108MotionBG/Shutterstock.com; p. 15 Arterra/Universal Images Group/Getty Images; p. 17 Gilbert S Grant/Science Source/Getty Images; p. 19 (cocoon) Josve05a/Wikimedia Commons; p. 19 (chrysalis) Daniele Della Mattia/Shutterstock.com; p. 21 (top left) Cbenjasuwan/Shutterstock.com; p. 21 (top right) Jason Yoder/Shutterstock.com; p. 21 (bottom left) Ricardo de Paula Ferreira/Shutterstock.com; p. 21 (bottom right) Sari ONeal/Shutterstock.com.

All rights reserved. No part of this book may be reproduced in any form without permission in writing from the publisher, except by a reviewer.

Printed in Canada

CONTENTS

Before Butterflies and Moths 4
Egg First 6
Larva Next 8
Pupa or Chrysalis? 18
All Grown Up 20
Glossary 22
For More Information 23
Index 24

Boldface words appear in the glossary.

Before Butterflies and Moths

The young of some animals look different from the adults. That's true of butterflies and moths. Before they're winged adults, butterflies and moths are caterpillars. Some people think caterpillars are gross. Do you? Read on before you decide!

Egg First

The lives of butterflies and moths have **stages**. Together, these stages form a **life cycle**. The first stage is an egg. A female butterfly or moth often lays her eggs on leaves or in the grass.

Larva Next

After some time, a larva comes out of the egg. It's called a caterpillar. It can look different depending on the **species** of butterfly or moth. Caterpillars come in many colors. They also have different numbers of legs!

No matter what kind of caterpillar it is, it's hungry. Caterpillars don't have teeth. They have two mouthparts called mandibles. Mandibles are sharp! The first thing a caterpillar eats is the covering of its own egg.

After its eggshell, the caterpillar starts to eat the plant it's on. Its egg is often laid on a plant the caterpillar likes to eat. Some people think caterpillars are pests because they eat plants. However, they're picky about food!

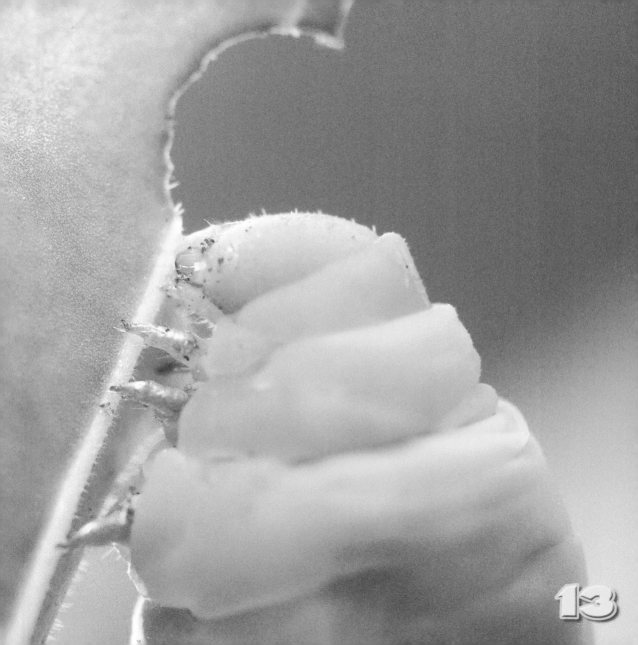

A caterpillar eats—a lot. It grows bigger as it eats. Soon, it doesn't fit in its skin. So, it **sheds** its skin! It **wiggles** out of the old skin. The soft skin underneath grows harder to fit the bigger body.

The caterpillar doesn't leave the old skin behind. It eats it! The skin, like the caterpillar's eggshell, has **nutrients** the caterpillar needs. A caterpillar may shed and eat its skin many times in just a few weeks!

Pupa or Chrysalis?

When a caterpillar is ready for its next life cycle stage, it hangs upside down. A moth spins a **cocoon** around its body. Now, it's called a pupa. A butterfly sheds its skin to uncover a hard outer shell called a chrysalis.

All Grown Up

The bug that comes out of the cocoon or chrysalis is an adult. Female butterflies and moths **mate** with males and lay eggs. Then, the life cycle starts all over again! Now, do you think caterpillars are gross—or cool?

21

GLOSSARY

cocoon: a covering which some bugs make around themselves to guard their body while they grow

life cycle: the series of stages through which a living thing passes from the beginning of its life until its death

mate: to come together to make babies

nutrient: something that plants, animals, and people need to live and grow

shed: to lose leaves, skin, fur, or feathers naturally

species: a group of animals or plants that are alike and can produce young animals or plants

stage: a point or period in the growth of something

wiggle: to move up and down or from side to side with short, quick motions

FOR MORE INFORMATION

BOOKS

Bodden, Valerie. *Moths*. Mankato, MN: Creative Education, 2014.

Hayes, Amy. *A Caterpillar Becomes a Butterfly*. New York, NY: Cavendish Square Publishing, 2016.

Marsh, Laura. *Caterpillar to Butterfly*. Washington, DC: National Geographic, 2012.

WEBSITES

Caterpillar
a-z-animals.com/animals/caterpillar/
Read more about this life cycle stage.

Life Cycle of Butterflies and Moths
www.kidsbutterfly.org/life-cycle
See a monarch butterfly grow from egg to adult.

Publisher's note to educators and parents: Our editors have carefully reviewed these websites to ensure that they are suitable for students. Many websites change frequently, however, and we cannot guarantee that a site's future contents will continue to meet our high standards of quality and educational value. Be advised that students should be closely supervised whenever they access the Internet.

INDEX

adults 4, 20
butterflies 4, 6, 8, 18, 20
chrysalis 18, 20
cocoon 18, 20
colors 8
eggs 6, 8, 10, 12, 20
food 12
grass 6
larva 8

leaves 6
life cycle 6, 18, 20
mandibles 10
moths 4, 6, 8, 18, 20
nutrients 16
plants 12
pupa 18
skin 14, 16, 18
species 8